WOOD
AND WOOD GRAINS

A PHOTOGRAPHIC ALBUM FOR ARTISTS
AND DESIGNERS

by

Phil Brodatz

DOVER PUBLICATIONS, INC.

NEW YORK

Published in Canada by General Publishing Company, Ltd., 30 Lesmill Road, Don Mills, Toronto, Ontario.
Published in the United Kingdom by Constable and Company, Ltd., 10 Orange Street, London WC 2.

Wood and Wood Grains: A Photographic Album for Artists and Designers is a new work, first published by Dover Publications, Inc., in 1971. Plates WW1 through WW109 are reproduced from original photographs by Phil Brodatz. The illustrations reproduced on Plates WW110 through WW112 are taken, with permission, from *Commercial Foreign Woods on the American Market* by David A. Kribs (rev. and enlarged edition, Dover Publications, Inc., 1968).

DOVER *Pictorial Archive* SERIES

Wood and Wood Grains belongs to the Dover Pictorial Archive Series. Up to three illustrations from this book may be reproduced on any one project or in any single publication free and without special permission. Wherever possible, please include a credit line indicating the title of this book, author and publisher. Please address the publisher for permission to make more extensive use of illustrations in this book than that authorized above.

The republication of this book in whole is prohibited.

The author can supply photographic prints of any of the illustrations in this book for $15.00 each. Please address order with remittance to: Phil Brodatz, 100 Edgewater Drive, Coral Gables, Florida 33133.

International Standard Book Number: 0-486-22424-4
Library of Congress Catalog Card Number: 71-140937

Manufactured in the United States of America
Dover Publications, Inc.
180 Varick Street
New York, N. Y. 10014

LIST OF PLATES

WOOD AND WOOD GRAINS

This book deals with wood.

I am involved in gathering a picture collection of textures, surfaces and interesting materials which show design in its broadest sense. This book deals with wood in relation to this interest only. We will look at new wood, old wood, cordwood, cross sections, wood in use, man-made fences, natural barriers, bark and some examples of the sculpture created by the forces of nature, time and weather.

Wood is a product of a living tree. It grows from a seed or as an outgrowth from a mother tree.

Each variety of tree has its own particular characteristics, influenced by soil, light, altitude, seasons and weather. Each individual tree has its own individual growth patterns. While two pieces of oak plank may look similar, nature with her infinite variations has decreed that two pieces can never be identical. We can carry this premise further, through all facets of nature and living things.

SECTION ONE   NEW WOOD (Plates WW1 to WW41)

The first section of the book examines new wood intimately for figure patterns. As a tree grows, it is subject to rain, wind, drought, insects. Each year the trunk thickens, and as it does it leaves a clear record of its progress. If we cut a trunk through we can read the annual rings. Each ring represents one year's growth, and there are good years, bad ones, some dry, some wet. These are pictured here by their size and color. An expert counting back can tell you what the weather was in the fall of 1895 in a certain locale. No two years are alike, no two trees are marked the same. No two pieces of wood can be identical twins. When the trunk is sawn conventionally, we see the rings as wood grain, beautiful patterns that can identify the wood. Much veneer is produced by shaving a trunk in an inward spiral fashion, thereby achieving another type of grain figure.

Therefore it is possible to produce quite different grain markings even from the same trunk, depending on the method of sawing.

Sometimes a growth appears on the side of a tree. When we cut this through we will find some very wild and irregular grain patterns, exceptionally interesting and beautiful in their irregularity. As in the human body, should some fault or damage occur, this area tends to become isolated by some very vigorous growth in the surrounding area. Many very fine pipes are carved from the burl of certain hardwood trees.

Redwood trees sprout from their burl sections.

SECTION TWO   OLD WOOD (Plates WW42 to WW63)

In searching for interesting patterns, we must look at old wood.

Old garden doors, shutters, very old beams, an ancient wharf and an old barge. Each of these could be a treatise, but for our purpose we can only consider the effects of "weathering" on the samples we have found. The fact that these examples are in excellent condition speaks well for the architect or builder who selected the particular wood with which to construct these things.

The wooden shutters (Plates WW60, WW62 and 63) are on a 300-year-old home in Florida. The rust-stained garden gate (WW58 and 59) is on a transplanted castle, very old and subjected to brilliant sunlight, salt spray and hurricanes.

The hand-hewn beams (WW48 and 49) are quite possibly over 200 years of age, removed and saved from an old building on the northern shore of Long Island. There are other buildings nearby which were constructed during the 1600's and it conceivably could be from one of these. I have read about the longevity and durability of the first-growth lumber that the early settlers had, and used with such abundance. They actually paved roads with logs for the early wagons.

The heavy dark timbers encrusted with sea life (WW44 and WW45) are on an old pier submerged in sea water when the tide was high. On the old barge (WW56 and WW57) even the rusted iron tie rods, and the weatherbeaten tongue and groove wood with the peeling paint, help make these textural patterns interesting.

As the tree grows there will be a variation of density or degree of hardness within each annual ring. The softer area of a wood grain will therefore be worn or weathered away first with a gradation to the denser wood. You may have noticed some new plywood walls with a real sculptured texture effect. This effect is manufactured by sandblasting out the softer areas of a veneer. Nature does the same thing to a plank, log or trunk, or root, and you will see some additional illustrations of this in the Nature's Sculpture section.

SECTION THREE   CROSS SECTIONS (Plates WW64 to WW73)

The end of a log dries rapidly when exposed to air. The rapid drying develops the radial check. I have seen some others, but the one shown in Plate WW64 is one of the most interesting that I have ever found.

Illustration WW64 was photographed in the studio with carefully controlled cross lighting. Compare this with WW65, which was photographed in the forest in northern California. Notice the similarity in the radial check, and the effect that the chain saw has on the texture.

In some illustrations this effect is so deep and interesting that the photograph concentrates on these marks alone (WW68 and WW69).

WW66 through WW73 are cross sections through the burl of an apple tree. This piece measured 30 inches in diameter, and it was a herculean task to cut two inch slices through the section. WW67 is a photograph at the farm, in daylight, of the section I left behind. The other photographs were taken at the studio, of the huge slices I dragged home.

The cross sections are a wonderful exercise in applied photography. Cross lighting showed up the chain saw marks. Flat, even copy lighting shows the beautiful wild grain patterns and the destroyed section with the unusual growth activity surrounding it. This wood also checked and cracked in drying. These cleavages added some extra interest to the total effect. Infrared film was used in photographing WW72 and WW73 to subordinate the surface complexion and inquire under the skin, and with this we have another design dimension. The use of this lighting and this film was the equivalent of peeling off layers of wood and probing deeper and deeper into each successive stage.

SECTION FOUR   FENCES (Plates WW74 to WW79)

When it became time for man to outline his land or enclose his possessions, he did this quite simply by using the materials he had at hand. In the open plain country he had lots of dry old sagebrush, and in the wooded areas he split rails. As he prospered he could manufacture lumber for fences. The beautiful and picturesque fences in WW74 and WW75 are used to enclose the thoroughbred horses on a farm in Lexington, Ky. In Kentucky these fences are traditionally painted white, while in Florida their counterparts are painted black.

SECTION FIVE   CORDWOOD (Plates WW80 to WW82)

I passed by a number of cordwood stacks, passed because they seemed too high, too regular, too commercial. There is a certain charm and quaintness to this old stack of wood. Yet it seems as though the wood cutter has just stepped away for a moment, and will shortly return to his work.

SECTION SIX   BARK (Plates WW83 to WW91)

This section contains a sampling of the packaging nature puts on her trees. Bark also has characteristics which are individual to each variety of tree.

SECTION SEVEN   NATURE'S SCULPTURE (Plates WW92 to WW109)

These are illustrations depicting nature as an artist. A master sculptor, using the trunks, branches or roots of trees as materials and time, rain and wind as tools, she fashions these fanciful creations. We must realize that the work is never done until it is all reduced to fragmentary soil.

We can look at these works and consider whence they have come and where they will go. In the twists and turns of simple debris we can find design and pattern, method and interest. Add light and shadow, and our work is in flux from sunup to sunset, and these objects may even be a sight to behold in moonlight too.

WW108 and WW109 are pictures of Monterey cypress at Point Lobos on the Pacific Coast. These trees spent a tortured lifetime clinging to these rocks on the Pacific palisades These trees were twisted, turned and pummeled by savage wind, rain and fog. And now after death they stand as silent statues, still defying the violent forces of nature.

SECTION EIGHT   PHOTOMICROGRAPHS (Plates WW110 to WW 112)

The final section consists of three plates with unusual views of nine different commercial woods. In each case the views show, from left to right, a photograph of the smooth cross section magnified ten times; (2) a photomicrograph of the cross section magnified 80 times; and (3) a photomicrograph of the tangential section magnified 100 times. These illustrations are reprinted with permission from *Commercial Foreign Woods on the American Market* by David A. Kribs (revised and enlarged edition published by Dover Publications, Inc., in 1968).

To be in tune with the world about us is to be constantly and painfully aware of the imminent possibility of total destruction. We face each day with a feeling that it may be the last. To preserve my hope and faith I turn to nature to find the true pattern of life and tempo of living, stability, dependability and the strength to survive.

I see everywhere manifestations of nature's master plan, the interdependence of living things, acclimated so well to their particular environment and to each other.

Scientifically, I see the organization and interdependence of land, sea, and air; the relationship of tiny vegetable plankton to the kingdom of the sea; the relationships of fish, birds and man to each other.

On land the simple presence of chlorophyll in living vegetable matter makes our existence physically possible. We use oxygen and expire carbon dioxide. Plants use our carbon dioxide and expire oxygen and water. Round and round goes this process.

I love to dig in the warm moist earth. Soil is a living, breathing reserve for life. It is composed of organic nitrogenous materials, soluble inorganic salts and insoluble minerals, waiting to be released by a catalyst; resting spores and seeds waiting for the moment to spring to life; worms, insects, disease organisms, and the antitoxins and antibodies to hold them in check. All is there so that it can generate and feed plants that will provide us with food and shelter for ourselves and the animals we need.

Whether this organization or balance of elements holding check rein on each other was derived through evolution or God-granted is not as important as the knowledge that it exists. To the student it is a beautiful plan, embracing total earth, the sun, the lunar cycle, the sea, land and air, infinite to the tiniest detail.

I need only to look at a green blade of grass to be reminded of its presence.

As certain as we are that the sun will set and rise again the next day, low tide follows high tide, and spring follows winter, birds fly south and return in the spring. From all this we can draw our strength and assurance that no matter what man does to man, nature will continue to provide our real basic necessities to live and survive.

When you see nature in her colorful raiment and in her many moods, you somehow feel all is right with the day and the world.

I wish to thank the people at the Georgia Pacific Corporation, the U. S. Plywood Corporation and the Union Lumber Company for their generous giving of their time and materials.

A special thanks to Hertha and Martin Lindblad, and Walter Wolfsohn, for their encouragement and very valuable assistance.

And a thank you to Mr. Hayward Cirker and his wonderful staff at Dover Publications for helping me over the rough spots, and for the final polish of the book.

                                                            PHIL BRODATZ

WW1   Chestnut.

WW2   Mahogany.

WW5   Sugar pine.

WW6   Sapeli.

WW7   White birch.

WW8   East Indian rosewood.

WW9   Teak.

WW10   Oak.

WW11   Mountain tulip.

WW12   Tigerwood.

WW13   Rift oak.

WW14   Sycamore.

WW15   Flat cut walnut.

WW16   Red birch.

WW19   Oak.

WW20   Ash.

WW21   Ash.

WW22   Mahogany burl.

WW23   Mahogany burl.

WW24   Walnut.

WW25   Walnut.

WW26   White pine.

WW27   White pine.

WW28   Quartered walnut.

WW29   Zebrawood.

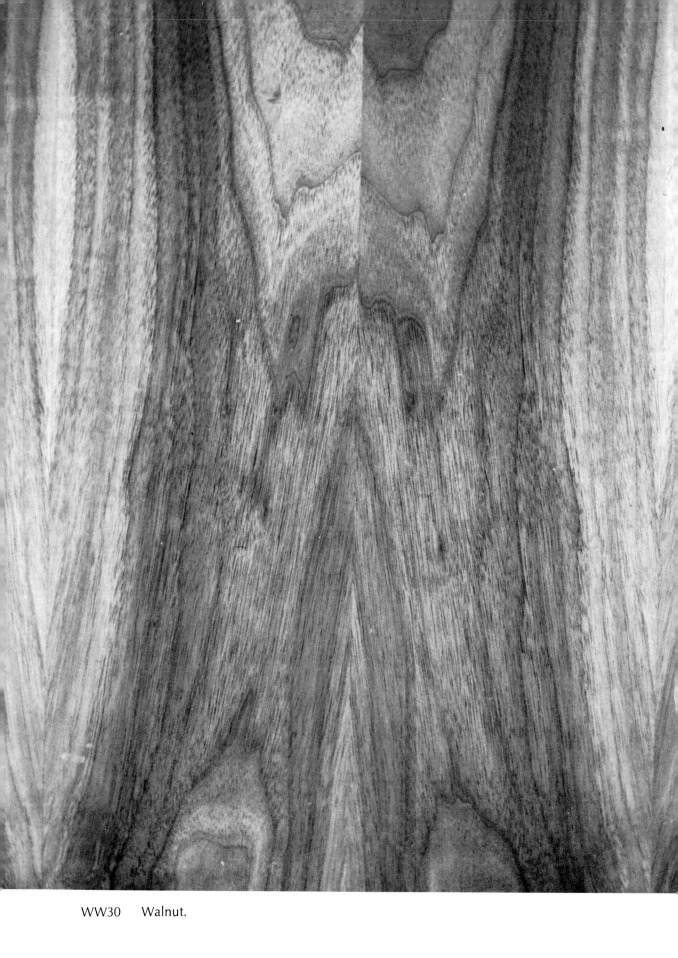

WW30   Walnut.

WW31   Maple.

WW32   Sapeli.

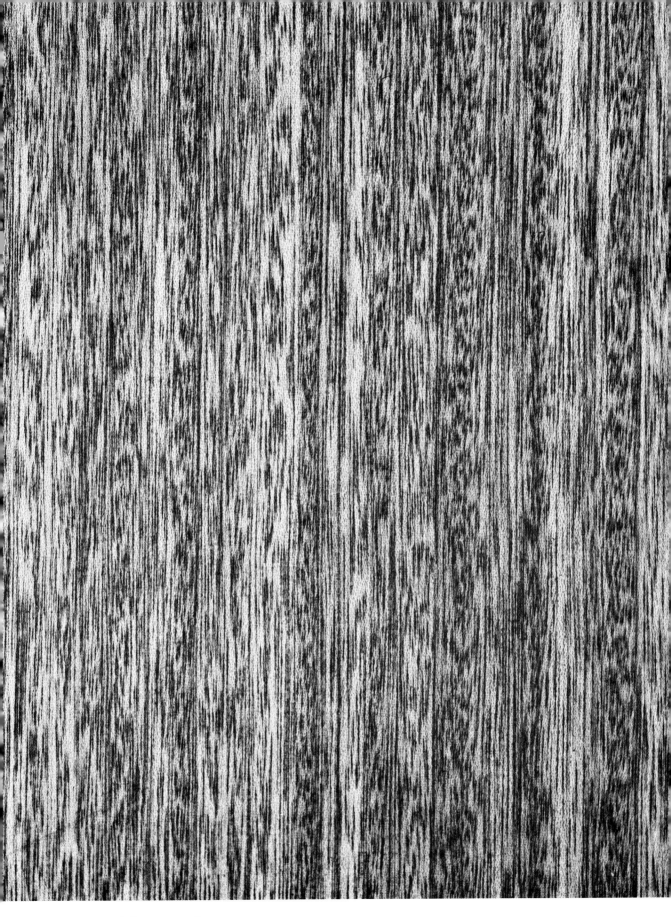

WW33   Philippine mahogany.

WW34   Birch.

WW35   Birch.

WW36   Cedar.

WW37   Mahogany.

WW38   Teak.

WW39   Mahogany.

WW40   Teak.

WW41   Satinwood.

WW42   Rough sawn redwood shiplap siding.

Even lighting. Reduced to ½ actual size.

WW43   Rough sawn redwood shiplap siding.

Cross lighting. Reduced to ½ actual size.

WW44   Old yellow pine pier underwater at high tide.

Actual size.

WW45   Old yellow pine pier underwater at high tide.

Actual size.

WW46   Pine boardwalk on an old pier, showing age and wear.

WW47   Pine boardwalk on an old pier.

WW48   Very old hand-hewn oak beams removed from an old farm building
        and saved. Note axe marks.

Approximately actual size.

WW49   Hand-hewn oak beams from an old farm building.

Approximately actual size.

WW50   Hand-hewn oak beams from an old farm building.

Approximately actual size.

WW51   Hand-hewn oak beams from an old farm building.

Approximately actual size.

WW52   Spruce.

WW53   Oak.

WW54   Old oak hand hewn beams.

WW55   Old oak hand hewn beams.

WW56   Deckhouse of an ancient barge at Oyster Bay, Long Island.

WW57   Side of an ancient barge at Oyster Bay, Long Island.

WW58   Close-up of a yellow pine garden gate in Florida.

WW59   Close-up of a yellow pine garden gate in Florida.

WW60　Closed wooden shutter on a stone building in the subtropics.

WW61   Detail of a garden doorway in the tropics.

WW62   The shutters (of WW60) open. Note the weathering of the pine
wood and the type of nails used.

WW63   The shutters (of WW60) open.

WW64   Cross section of log, showing radial check and saw tooth marks.

Studio photograph. Magnification 1½ x.

WW65   Cross section of log in a California forest. Notice radial check.

Photographed in daylight. Reduced to ½ actual size.

WW66   Cross section of apple tree burl.

WW67   Cross section of apple tree burl.

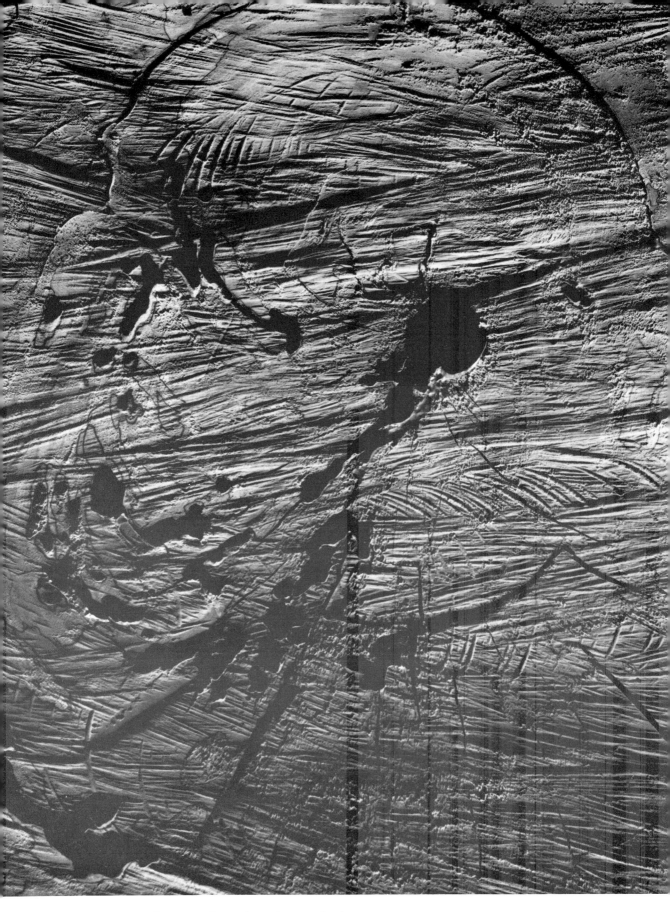

WW68　Cross section of apple tree burl.

Cross lighting to accent chain sawtooth marks.

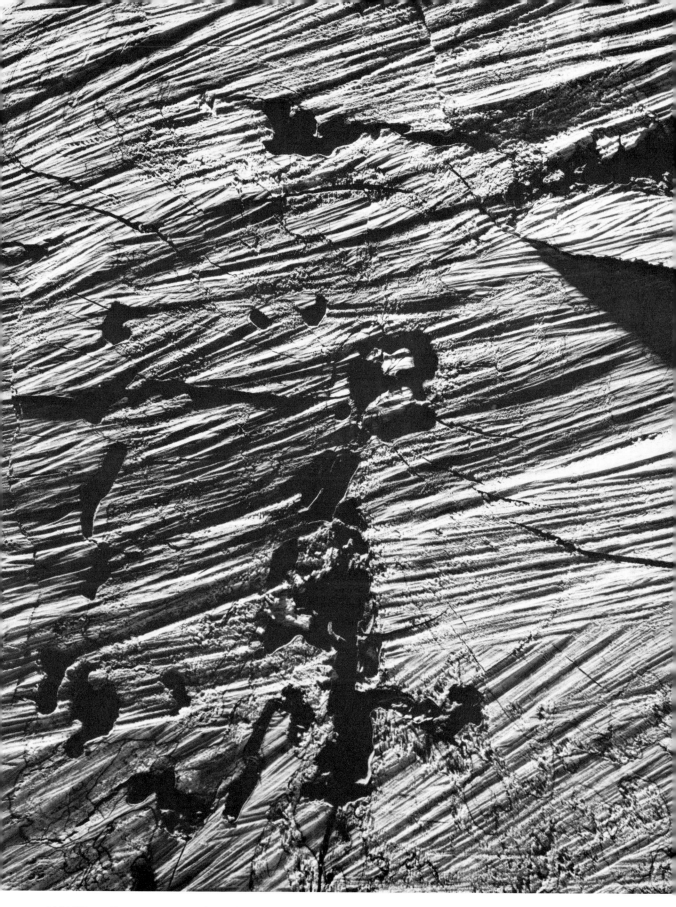

WW69   Cross section of apple tree burl.

Cross lighting to accent chain sawtooth marks.

WW70   Cross section of apple tree burl.

Even copy lighting to show maximum detail.

WW71   Cross section of apple tree burl.

Even copy lighting to show maximum detail.

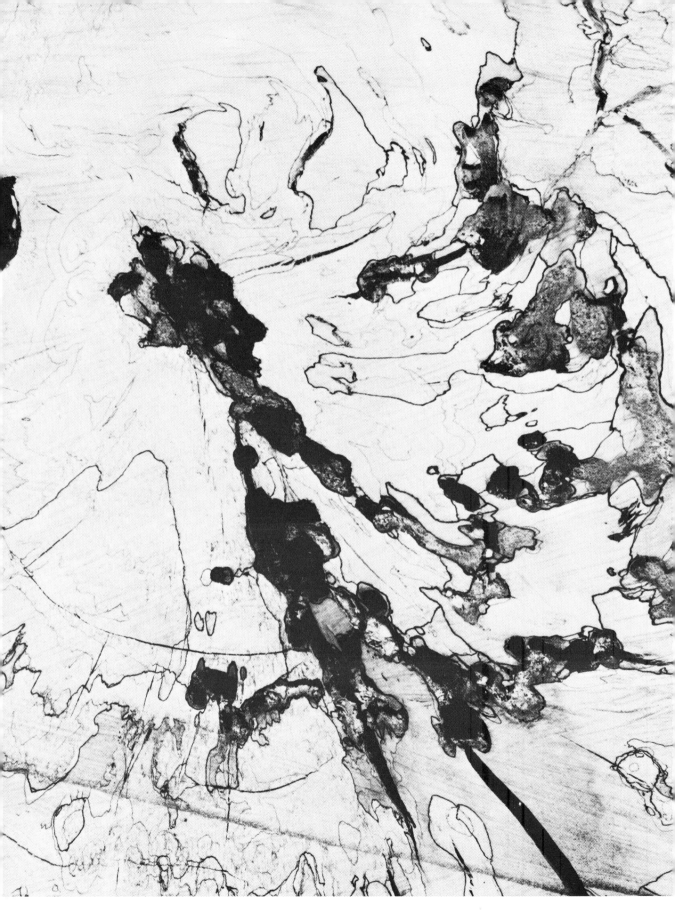

WW72   Cross section of apple tree burl.

Infrared film with A25 filter to eliminate the gray tones and to accent the
harder wood lines.

WW73   Cross section of apple tree burl.

Infrared film with A25 filter.

WW74   Man-made fence on a horse farm in Lexington, Ky.

WW75   Man-made fence on a horse farm in Lexington, Ky.

WW76   Sagebrush to fence cattle in Nevada.

WW77   Natural split rail for cattle fencing in California.

WW78   Bulkheading a river bank on Long Island.

WW79   Lattice-type fence for drifting sand and snow on an Atlantic Ocean beach.

WW80   Cordwood pile.

WW81   Cordwood pile.

WW82   Cordwood pile.

WW83   Melaleuca leucadendron.

WW84   Melaleuca leucadendron.

WW85   Tupelo.

WW86   White ash.

WW89   Palm.

WW90   Black locust.

WW91   Redwood.

WW92   Nature's sculpture (I).

WW93   Nature's sculpture (II).

WW94   Nature's sculpture (III).

WW95   Nature's sculpture (IV).

WW96   Nature's sculpture (V).

WW97   Nature's sculpture (VI).

WW98   Nature's sculpture (VII).

WW99   Nature's sculpture (VIII).

WW100   Nature's sculpture (IX).

WW101   Nature's sculpture (X).

WW102   Nature's sculpture (XI).

WW103   Nature's sculpture (XII).

WW104   Nature's sculpture (XIII).

WW105   Nature's sculpture (XIV).

WW106   Nature's sculpture (XV).

WW107   Nature's sculpture (XVI).

WW108   Nature's sculpture (XVII): Monterey cypress at Point Lobos, Calif.

INDEX